Let's Look at Lizards

GECKOS

By Seth Lynch

Gareth Stevens
PUBLISHING

Please visit our website, www.garethstevens.com. For a free color catalog of all our high-quality books, call toll free 1-800-542-2595 or fax 1-877-542-2596.

Library of Congress Cataloging-in-Publication Data
Names: Lynch, Seth, author.
Title: Geckos / Seth Lynch.
Description: Buffalo, New York : Gareth Stevens Publishing, [2025]. | Series: Let's look at lizards! | Includes bibliographical references and index.
Identifiers: LCCN 2023034773 | ISBN 9781538292709 (library binding) | ISBN 9781538292693 (paperback) | ISBN 9781538292716 (ebook)
Subjects: LCSH: Geckos–Juvenile literature.
Classification: LCC QL666.L245 L96 2025 | DDC 597.95/2–dc23/eng/20230728
LC record available at https://lccn.loc.gov/2023034773

First Edition

Published in 2025 by
Gareth Stevens Publishing
2544 Clinton Street
Buffalo, NY 14224

Editor: Kristen Nelson
Designer: Leslie Taylor

Photo credits: Cover Kurit afshen/Shutterstock.com; p. 5 MZPHOTO.CZ/Shutterstock.com; p. 7 BLFootage/Shutterstock.com; p. 9 (left) Artush/Shutterstock.com, (right) Ondrej Prosicky/Shutterstock.com; p. 11 N8Allen/Shutterstock.com; p. 13 (green) Fedor Selivanov/Shutterstock.com, (yellow) litag22/Shutterstock.com, (pink) Kathryn Sullivan/Shutterstock.com; p. 15 Kurit afshen/Shutterstock.com; p. 17 Matt Jeppson/Shutterstock.com; p. 19 Brittany Mason/Shutterstock.com; p. 21 ePhotocorp/iStockphoto.com; p. 23 lessysebastian/Shutterstock.com.

Printed in the United States of America

CPSIA compliance information: Batch #CSGS25: For further information contact Gareth Stevens, New York, New York at 1-800-542-2595.

Contents

Geckos are lizards.
They make a sound.
It's like a bark!

They can climb. Tiny hairs on their feet make them stick!

They love sunny places.
They live in deserts.
They live in rainforests.

Some geckos are long. Others are as small as a thumb!

They can be lots of colors!
They can be gray or green.
They can be yellow or pink.

They eat bugs.
They eat mice.

Their tails can fall off.
It keeps them safe.

They are awake
at night.

Mothers lay one
or two eggs.
They take care
of their eggs.

Babies grow fast.
They look like adults.

Words to Know

desert

egg

rainforest

Index